A TROOP OF TARANTULAS

BY REBECCA STORM

CONTENTS

Copyright © 2025 Hungry Tomato Ltd

First published in 2025 by Hungry Tomato Ltd
F15, Old Bakery Studios, Blewetts Wharf, Malpas Road, Truro, Cornwall,
TR1 1QH, UK.

A CIP catalogue record for this book is available from the British Library.

ISBN 9781835694176

Printed in China

Discover more at
www.hungrytomato.com

Picture credits:
Abbreviations: m-middle, t-top, l-left, r-right, bg-background.

Alamy: 8b, 32tr. FLPA: 4ml, 4br, 5b, 9tr, 12ml, 18b, 19r. Science Photo Library: 14m, 17b. OSF: 21b. Shutterstock: Amar Dhani 10b; common human 25br; douglas cliff 27tr; I Wayan Sumatika FC; linn currie 27tr; macronatura.es 1bg, 22b; Mark Breck 9b; Pets in Frames 6-7bg, 31br; Milan Zygmunt 23tr, 28bl, 29tr; Samurary 13br; Serpeblu 13t; Stephanie Budouze 29ml; Rares_Reptilis 15t; wanchat M 4tl, 15b; Wirestock Creators 17tr.

Every effort has been made to trace the copyright holders, and we apologise in advance for any unintentional omissions. We would be pleased to insert the appropriate acknowledgements in any subsequent edition of this publication.

DISCLAIMER:
Spiders are fascinating, but best to stay away!
Don't touch or handle them – some can bite or
get aggressive when they feel threatened.

Words in **BOLD** can be found in the glossary.

TARANTULAS

Tarantulas are spiders. Spiders are wingless, eight-legged animals. They are not **insects**, because insects only have six legs. Tarantulas are the largest of all spiders!

HOW DO THEY LIVE?

Like all spiders, tarantulas are **predators**. They are **carnivores** that hunt other animals. Spiders mainly eat insects and other bugs, but tarantulas will sometimes hunt much bigger **prey**, such as birds.

Mexican redknee tarantula

WHERE DO THEY LIVE?

Spiders are able to live anywhere, except the North and South Poles, and on the tops of mountains. Tarantulas like to live in warm places, such as deserts and rainforests.

A tarantula exploring the rainforest floor

UNDERSTANDING THE BUGS

Spiders belong to a group of creatures known as **arachnids**. Arachnids, like insects, are part of a larger group of animals known as **arthropods**. Arthropods do not have an inner **skeleton** made of bones, like humans do. Instead, they have a tough outer skin called an **exoskeleton**.

To grow, tarantulas form a new exoskeleton and shed the smaller one!

PARTS OF A TARANTULA

A large tarantula has a body that is about 13 cm (5 inches) across, with eight legs. Each leg can measure up to 28 cm (11 inches) in length. Most tarantulas are covered in lots of hairs.

All spiders have two parts to their body: the **prosoma** and the **abdomen**. The front part of the prosoma is the spider's head – they have eyes, a brain, strong **jaws**, and sharp **fangs**.

Prosoma

Towards the front of the prosoma is a pair of short **pedipalps** (arms) that are used for holding food close to their jaws.

SPIDER SHAPE

The earliest spiders probably had three parts to their bodies: head, **thorax**, and abdomen, just like insects. Over time, the spider's head and thorax joined together into a single part, called the prosoma. With insects, however, the head and the thorax stayed separate.

Brazilian pink tarantula

Inside the abdomen, tarantulas produce **silk**.

Behind the pedipalps, there are four pairs of walking legs.

HOME SWEET HOME

Tarantulas live by themselves and do not form family groups. They can live very long lives. Females can live to be over 30 years old, while males only live to around half that age.

Tarantulas like to live in safe, dry, stable nests. Some may live in the same nest for the whole of its adult life.

Some tarantulas live in holes in trees, but most prefer burrows underground.

A tarantula digs its own burrow, which can be as far as 76 cm (30 inches) below the ground. Using its strong jaws, it can easily cut through even the hardest soil. The opening of the burrow is kept very small so that no predators can fit through.

Tarantula with a small burrow opening, to stop other creatures getting in.

SILK PRODUCTION

Spiders, and some insects, can produce silk. With insects, it is the **larvae** that produce silk. They spin a **cocoon** of silk around their bodies while growing. With spiders, however, it is the adults that spin silk. Spiders use silk for building **webs**, making trip lines, and lining their nests.

A spider in its silk web

SPINNING WEBS

Most tarantulas spend a lot of their long lives hiding in their burrows and waiting for prey to come along. After a big meal, a tarantula does not need to eat again for several weeks!

Some spiders spin webs to catch flying insects, or to trap insects that crawl along the ground, but tarantulas do something different. They use their silk for protecting their burrows.

A tarantula will often spin trip lines that spread out from its burrow. The silk is thin, but very strong, making it perfect to protect their nest.

A huntsman spider catches a fly

One end of the silk is attached to something solid, while the other is attached to the tarantula's silk nest. A tarantula will feel when an animal touches the line, rushing out quickly to attack its prey!

PARALYSING VENOM

All spiders produce **venom**, which is made by special **glands** and injected into a victim through sharp fangs in the spider's jaw. The venom is used to stop prey from moving. Spiders prefer their prey to be alive when they start eating!

Wolf spider

PROWLING FOR PREY

Some tarantulas do not like waiting for their prey, so they regularly go out and hunt. The largest, bird-eating tarantulas are an example of spiders that like to hunt a lot!

A tarantula out in daytime

It is usually dangerous for tarantulas to hunt in daylight. Their large size makes them an easy target for spider-eating birds that may be flying above them. By night, it is much safer.

Tarantulas live mainly on a diet of large insects. However, the biggest tarantulas are able to attack small **mammals**, lizards, and even birds that are sitting in their nests.

If a tarantula cannot find any large prey, it will collect lots of small insects. The spider will then wrap them together with strands of silk and eat them all at the same time!

A spider attacking large prey

LIQUID STEEL

Spider silk starts out as a liquid, made by special glands inside the abdomen. The liquid is squirted out of the spider's body by tiny nozzles called spinnerets. The spider uses its legs to pull out the strands of liquid silk until they harden. Spider silk is incredibly strong!

A spider's spinneret, where its silk comes from

SPIDER SENSES

Some spiders have excellent eyesight, but tarantulas don't. However, all spiders can feel vibrations, and they use them to get information about their surroundings.

PAIRED EYES

Most spiders have eight eyes, although some have six, four, or only two. Usually, the eyes are arranged in two rows. Some spiders, such as jumping spiders, have two eyes that are bigger than the others, which produce really clear images. Most spiders, however, like tarantulas, have to make do with poor eyesight!

The eyes of a jumping spider

Tarantulas have eight small eyes, but each eye can do little more than tell the difference between light and darkness. Tarantulas can just about see movement through them.

A spider's leg hairs

All spiders have lots of hairs for feeling **vibrations**. Tarantulas can feel tiny vibrations moving below each of their eight legs. Because their legs are spread out, they can also tell which direction the vibrations are coming from.

IMPRESSIVE JAWS

There are two main groups of spiders, according to which way their jaws work. Tarantulas belong to the group that have jaws that strike downwards.

A tarantula's sharp fangs

All tarantulas have two sharp fangs that they use to stop their prey from moving. They can bring them down at the same time, or one after the other.

LIQUEFIED FOOD

Spiders do not chew and crunch their food – they slurp it up! Once their prey stops moving, the spider dribbles digestive juices over their prey. These juices turn the prey into semi-liquid goo that the spider sucks into its mouth. Gross!

The other group of spiders have jaws that strike inwards. Their fangs can be moved close together and used like a pair of pincers. These fangs are better than the downwards-striking jaws because they stop prey getting away!

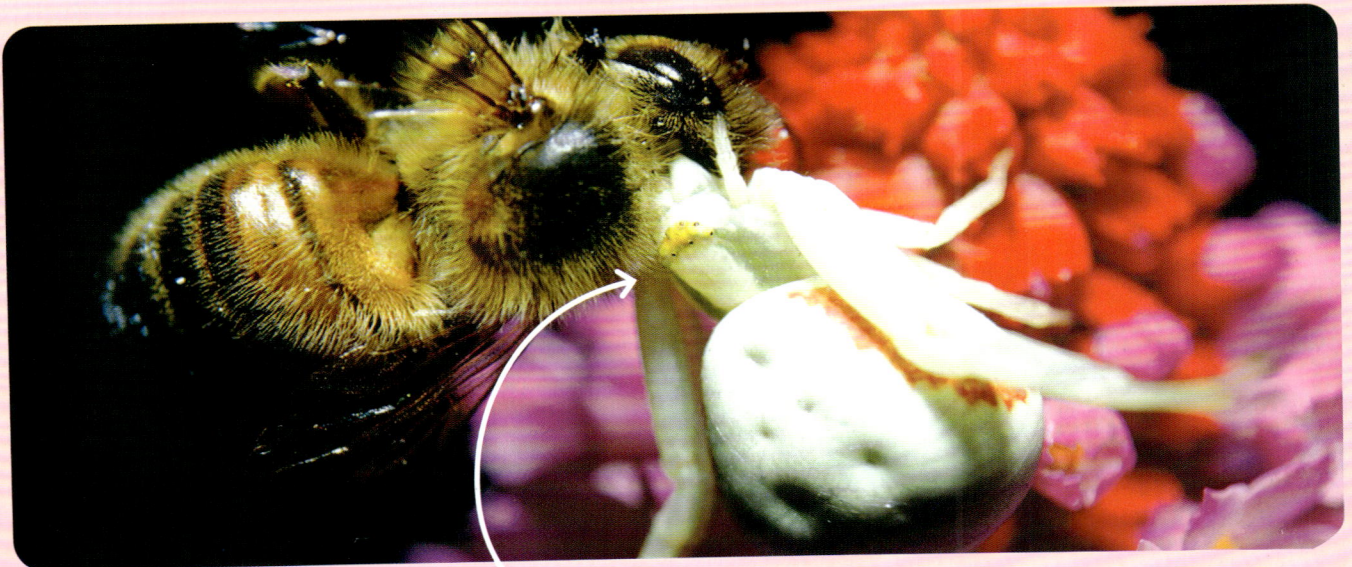

A crab spider sinking its fangs into a honeybee

LOOKING FOR A MATE

Tarantulas develop very slowly and are not able to mate until they are about 10 years old. Mating normally takes place during the hottest and driest part of the year.

Males and female tarantulas live separately, so in order to find a mate, a male tarantula must go out looking for a female.

A female California tarantula is alerted to a male outside her nest, thanks to her trip lines.

Mating can be a risky business for males because female tarantulas often prefer eating to mating – and that includes eating male tarantulas!

When a male tarantula finds a female to mate with, it doesn't mean he is safe. After mating, the female tarantula may still decide to eat him!

SAFETY SOUNDS

Spiders can make sound in lots of different ways, such as by rubbing their legs together or drumming their legs on the ground. Many spiders use these sounds to communicate.

Male tarantulas in search of a mate make a special safety sound when they are on the move. If they make this sound, they are less likely to be attacked (and eaten!) by a female.

This jumping spider makes sure it is safe for him to mate with the female.

BABY SPIDERLINGS

Female tarantulas can lay around 2,000 eggs at a time. Each egg is very small, so the female holds them together with strands of silk. This is called an egg sac.

This ball of silk and eggs takes up around half the size of the female's body. The sticky silk stops eggs from falling from the egg sac.

The female usually leaves the egg sac in the deepest part of her burrow and does not give the eggs much attention. However, if it rains and the burrow floods, the female will carry the egg sac to safety.

Female wolf spiders also carry egg sacs to and from their nests.

The eggs hatch into tiny **spiderlings** that look just like miniature adults. The spiderlings feed on tiny bugs in the soil that are much too small to fill up an adult tarantula.

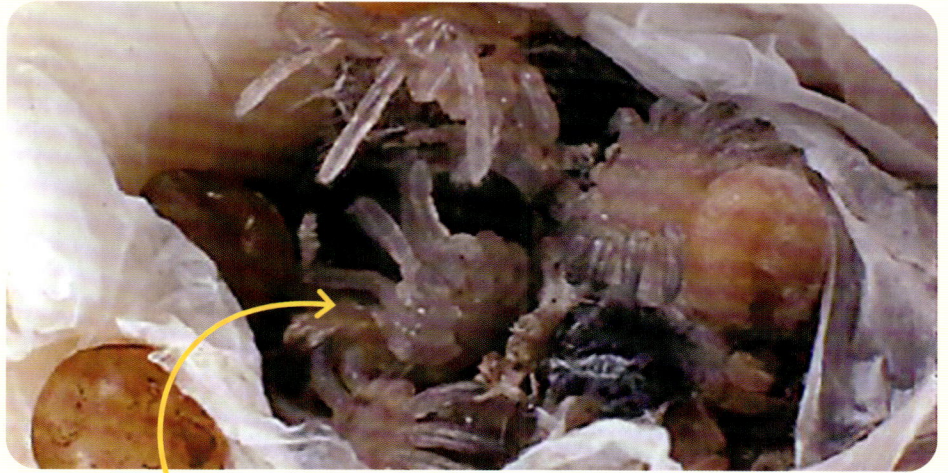

Tiny spiderlings hatching out of the egg sac

SPIDER DEVELOPMENT

A spider's exoskeleton is not stretchy – it does not budge as the spider grows. Instead, a growing spiderling develops a new exoskeleton beneath the old, outgrown one. When the new exoskeleton is ready, the old one splits open and is discarded. This process is known as **moulting**.

A tarantula moulting

HUMANS AND TARANTULAS

The spiders that we call tarantulas, even the biggest ones, are not so dangerous to humans. Although their powerful jaws can give a painful bite, the venom of tarantulas is too weak to have much effect on us.

The name "tarantula" originally belonged to a much smaller, and much more dangerous type of spider. This original tarantula is what we know now as a European wolf spider. The wolf spider was first named "tarantula" after the town of Taranto in Southern Italy, where it lives.

Scientists discovered that, although the wolf spider gives a painful bite, it does not give humans any sort of disease. Because the original tarantula was actually a wolf spider, the name "tarantula" has since been given to the big hairy spiders we know today!

Goliath birdeater tarantula

IRRITATING HAIRS

As well as their sharp fangs and venom, some tarantulas have another deadly feature – their hair! Their bodies are covered in hairs that are coated with poison, and have sharp, fragile tips that easily break off.

Golden knee tarantula

ALL SORTS OF SPIDERS

There are more than 40,000 different species of spider. They all have two parts to their bodies and eight legs, but some spiders look very different to others!

MARCRACANTHA

Some small spiders have spines and horns, giving them an unusual appearance. Scientists believe that some spiders developed horns and spines to make it difficult for birds to swallow them!

FLOWER SPIDER

These spiders are also called crab spiders, as they often walk sideways, just like crabs do! They wait on flowers to catch visiting insects. Some flower spiders can change the colour of their bodies to match the flowers they are sitting on!

BLACK WIDOW

The black widow is easy to spot by the bright red markings on its body. This small creature has sharp fangs that inject venom. The bite of a black widow is very painful, and the venom can be deadly, even for humans!

JUMPING SPIDER

This type of spider is one of the largest of the spider groups. All jumping spiders have very good eyesight, and they have two pairs of eyes they can see from. They use their good eyesight to jump down on unsuspecting prey!

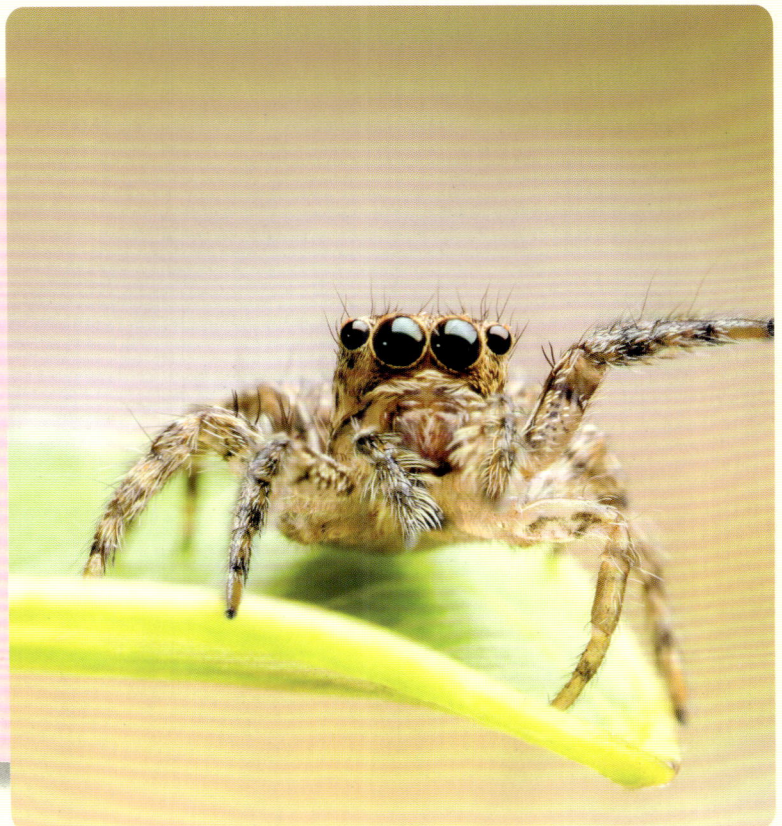

SPIDER SILK TACTICS

Many spiders live and hunt in much the same way, but others have different behaviours, and very different ways of using their silk.

WEB SPINNERS

The most well-known spiders are those that spin webs between branches to catch flying insects. These spiders sit at one corner of their web and wait for vibrations made by insects stuck in the web.

WATER SPIDER

This spider is only found in fresh water. It spins a web of silk, and then traps bubbles of air in the web. The air in the bubbles allows the spider to breathe underwater while it waits for prey, such as **tadpoles** and small fish.

NET-CASTING SPIDER

Net-casting spiders choose to spin small webs, similar to fishing nets. The spider then holds this net between its front legs and waits patiently on a branch. When an insect comes within reach, the spider drops the net to catch its prey!

SPITTING SPIDER

Spitting spiders have special venom glands that allow them to spit two streams of poisonous silk. They attack their prey by covering them with a zigzag pattern of threads so they can't escape!

FUN TARANTULA FACTS

There's so much more to know about tarantulas! Delve into some fantastic facts about these creepy critters.

TARANTULAS CAN BE...

as small as a fingernail or as big as a dinner plate!

TARANTULAS HAVE...

retractable claws, just like cats!

SOME TARANTULAS...

eat lizards, birds, and even mice!

FOR MOST PEOPLE...

tarantula bites are no worse than a bee sting.

A TARANTULA SPIDERLING...

will moult about 10 times before it becomes a fully grown adult.

TARANTULAS LIKE TO...

live alone, and will attack other tarantulas that come near them uninvited.

A FEAR OF SPIDERS IS...

called **arachnophobia**. It is one of the most common fears among humans.

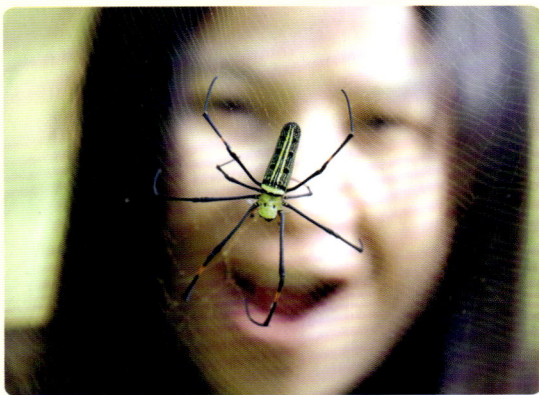

THERE ARE MORE THAN...

800 species of tarantulas.

WHEN FEELING THREATENED, THEY...

can make a loud hissing noise by rubbing the bristles on their legs together.

GLOSSARY

Abdomen – the largest part of a spider's body: the abdomen, contains most of the important organs.

Arachnids – a group of bugs that includes spiders, scorpions, ticks, and mites.

Arachnophobia – the fear of spiders.

Arthropods – bugs that have jointed legs; insects and spiders are arthropods.

Bug – one of a large number of small land animals that do not have a skeleton.

Carnivores – animals that eat meat.

Cocoon – a protective covering of silk produced by insect larvae (see right) to protect their bodies while they transform into adults.

Exoskeleton – a hard outer covering that protects and supports the bodies of many bugs.

Fangs – long, sharp teeth. Some fangs are designed to inject venom (see right).

Glands – a part of an animal's body that is used to make particular substances, such as silk.

Insect – a type of very small animal with six legs, a body divided into three parts, and usually two pairs of wings.

Jaws – hinged structures around the mouth that allow some animals to bite and chew.

Larvae – grub-like creatures that are in the juvenile (young) stages in the life cycle of many insects.

Mammals – a group of warm-blooded animals that have an internal skeleton (see right), and which feed their young with milk.

Mate – one of a pair of animals that live or have babies together.

Moulting – the shedding of a spider's exoskeleton (see left), which they need to do in order to grow.

Pedipalps – short, leg-like organs that a spider uses to hold its food.

Predators – animals that hunt and eat other animals.

Prey – an animal that is eaten by other animals.

Prosoma – the front part of a spider's body that consists of the head and the thorax, fused together into a single body part.

Retractable – something that can be taken or drawn back.

Silk – a natural thread produced by insect larvae (see left) and adult spiders.

Skeleton – an internal structure of bones that support the bodies of large animals such as mammals, reptiles, and fish.

Species – a group of living things that share characteristics and features.

Spiderlings – young spiders that are not yet fully grown.

Spinnerets – tiny nozzles on a spider's abdomen (see left) that are used to squirt out silk.

Tadpoles – the juvenile (young) form of frogs or toads.

Thorax – the middle part of an insect's body where the legs are attached.

Venom – a poison produced by an animal for use against other animals.

Vibrations – to move back and forth very quickly.

Web – a network of silk threads produced by many spiders to catch insects.

INDEX